Have You Seen An Elephant's Nest?

Text copyright © 1993 by The Child's World, Inc.
All rights reserved. No part of this book may be
reproduced or utilized in any form or by any means
without written permission from the Publisher.
Printed in the United States of America.

Distributed to Schools and Libraries
in Canada by
SAUNDERS BOOK COMPANY
Box 308
Collingwood, Ontario, Canada 69Y3Z7 / (800) 461-9120

Library of Congress Cataloging-in-Publication Data
Woodworth, Viki.
Have you seen an elephant's nest? / Viki Woodworth.
p. cm.
Summary: Illustrations and brief rhyming text
ask which animal lives in a nest, a den, a burrow,
or the hollow of a tree.
ISBN 0-89565-824-0
1. Animals – Habitations – Juvenile literature.
[1. Animals – Habitations.] I. Title.
OL756.W64 1993   91-37637
591.56'4–dc20   CIP / AC

Have You Seen An Elephant's Nest?

by Viki Woodworth

Bloomfield Twp. Public Library

Viki Woodworth

Viki Woodworth graduated from Miami University in Oxford, Ohio. Though trained as an art teacher, she chose to write and illustrate children's books as a way to teach and reach children. She lives in Seattle, Washington with her husband and two young daughters.

A bird
or a glove?

(A Bird)

(A Bear)

(Squirrel)

What lives in burrows under your feet?

A hot dog?

A horn?

A mole

or a parakeet?

(A Mole)